S0-EGO-687

TOOLS FOR CAREGIVERS

- **ATOS:** 0.9
- **GRL:** C
- **WORD COUNT:** 44
- **CURRICULUM CONNECTIONS:** insects, nature

Skills to Teach

- **HIGH-FREQUENCY WORDS:** a, and, at, has, have, I, in, is, it, let's, look, one, see, they, this, up
- **CONTENT WORDS:** antennas, beetles, cases, firefly, ladybug, legs, lights, lives, six, spots, swims, two, water, wings
- **PUNCTUATION:** exclamation points, periods
- **WORD STUDY:** compound words (*firefly, ladybug*); long /e/, spelled *ee* (*beetles*); /oo/, spelled *oo* (*look*); /oo/, spelled *wo* (*two*); multisyllable word (*antennas*)
- **TEXT TYPE:** information report

Before Reading Activities

- Read the title and give a simple statement of the main idea.
- Have students "walk" though the book and talk about what they see in the pictures.
- Introduce new vocabulary by having students predict the first letter and locate the word in the text.
- Discuss any unfamiliar concepts that are in the text.

After Reading Activities

There are many kinds of beetles. All beetles have six legs, two antennas, wings, and wing cases. Ask the readers if they have seen any of the beetles in the book. Have they seen other kinds? What did they look like? Where did the readers seem them? Have each reader draw a beetle and label its parts.

Tadpole Books are published by Jump!, 5357 Penn Avenue South, Minneapolis, MN 55419, www.jumplibrary.com

Editor: Jenna Trnka **Designer:** Michelle Sonnek

Photo Credits: Cosmin Manci/Shutterstock, cover; Protasov AN/Shutterstock, 1; MirekKljewski/iStock, 3; MVolodymyr/Shutterstock, 2tl, 2mr, 4–5; Nastya22/Shutterstock, 2br, 6–7; Jeff Lepore/Alamy, 2bl, 8–9; irin-k/Shutterstock, 2ml, 10–11; Cathy Keifer/Shutterstock, 2tr, 12–13; John CleggPanthe/Pantheon/SuperStock, 14–15; Marco Uliana/Shutterstock, 16.

Library of Congress Cataloging-in-Publication Data
Names: Nilsen, Genevieve, author.
Title: I see beetles / by Genevieve Nilsen.
Description: Tadpole books edition. | Minneapolis, MN: Jump!, Inc., (2020) | Series: Backyard bugs | Audience: Age 3–6. | Includes index.
Identifiers: LCCN 2018050518 (print) | LCCN 2018051535 (ebook) | ISBN 9781641287944 (ebook) | ISBN 9781641287920 (hardcover: alk. paper) | ISBN 9781641287937 (paperback)
Subjects: LCSH: Beetles—Juvenile literature.
Classification: LCC QL576.2 (ebook) | LCC QL576.2 .N55 2020 (print) | DDC 595.76—dc23
LC record available at https://lccn.loc.gov/2018050518

ARD BUGS

I SEE BEETLES

by Genevieve Nilsen

TABLE OF CONTENTS

tadpole books

WORDS TO KNOW

antennas

firefly

ladybug

legs

wing cases

wings

I SEE BEETLES

Let's look at beetles!

They have six legs.

And two antennas.

wing

They have wings.

wing case

They have wing cases.

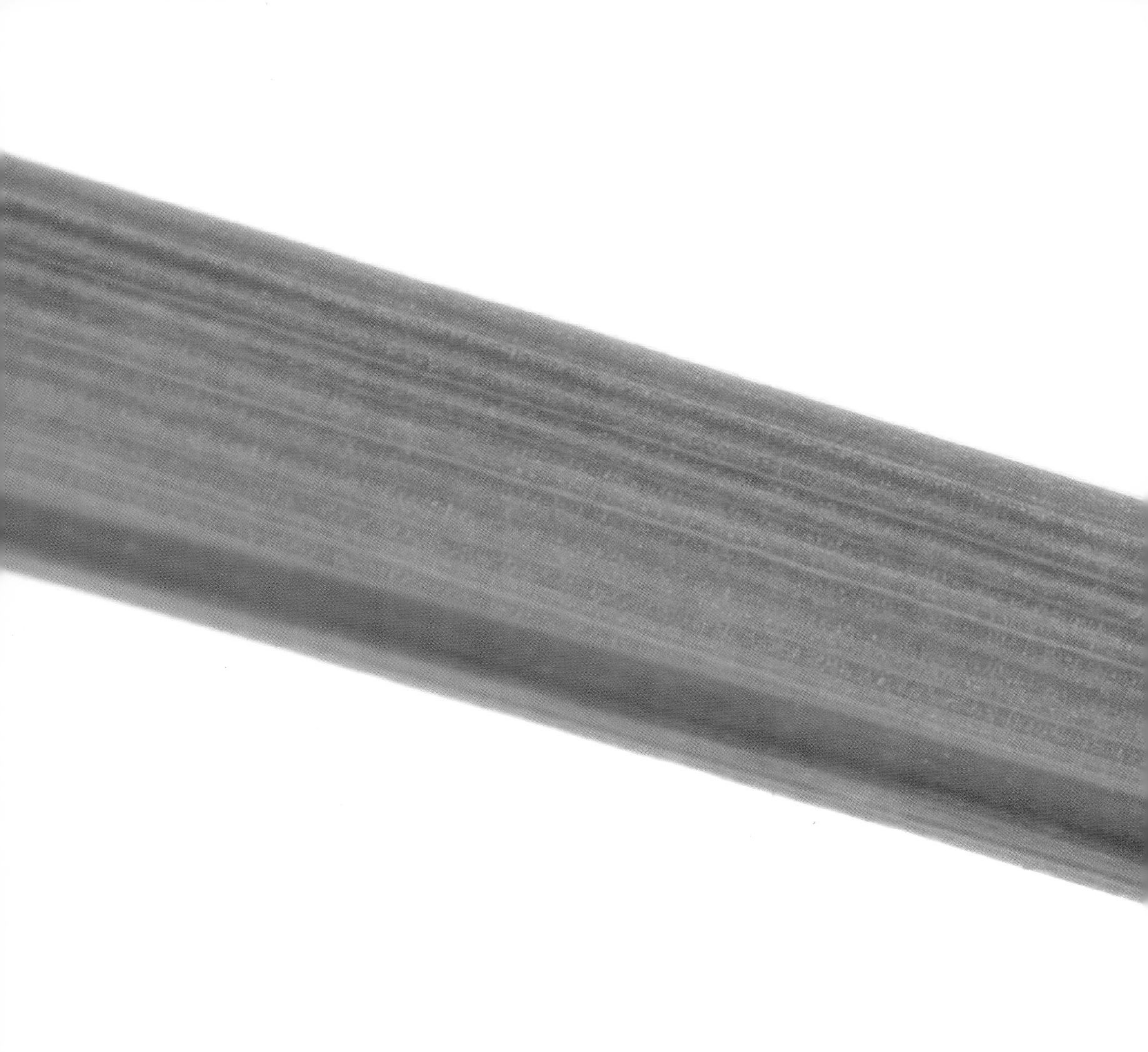

A ladybug is a beetle.

It has spots!

A firefly is a beetle.

It lights up!

This one swims.

It lives in water!

LET'S REVIEW!

This is a beetle! How do we know? It has six legs, antennas, and wing cases. Point to each.

INDEX